INS  NEERS

Managing and minimizing construction waste

A practical guide

J. Ferguson, OBE, N. Kermode, C. L. Nash,
W. A. J. Sketch and R. P. Huxford

Published by Thomas Telford Publications, Thomas Telford Services Ltd, 1 Heron Quay, London E14 4JD

First published 1995

Distributors for Thomas Telford books are
USA: American Society of Civil Engineers, Publications Sales Department, 345 East 47th Street, New York, NY 10017-2398
Japan: Maruzen Co. Ltd, Book Department, 310 Nihonbashi 2-chome, Chuo-ku, Tokyo 103
Australia: DA Books and Journals, 648 Whitehorse Road, Mitcham 3132, Victoria

A catalogue record for this book is available from the British Library

Classification
Availability: Unrestricted
Content: Guidance based on waste management law and best current practice
Status: Committee guided
User: Civil engineers, waste authorities

TD
793.95
.M3
1995

WIDENER UNIVERSITY
WOLFGRAM
LIBRARY
CHESTER, PA

DISCARDED BY WIDENER UNIVERSITY

ISBN: 0 7277 2023 6

© Institution of Civil Engineers, 1995

All rights, including translation reserved. Except for fair copying, no part of this publication may be reproduced, stored in a retrieval system or transmitted in any form or by any means, electronic, mechanical, photocopying or otherwise, without the prior written permission of the Publisher: Books, Publications Division, Thomas Telford Services Ltd, Thomas Telford House, 1 Heron Quay, London E14 4JD.

This book is published on the understanding that the author is solely responsible for the statements made and opinions expressed in it and that its publication does not necessarily imply that such statements and/or opinions are or reflect the views or opinions of the publishers or the Institution of Civil Enginerrs.

Typeset using Corel Ventura 4.2 at Thomas Telford Services Ltd
Printed in Great Britain by Ashford Press

Produced through the Association of Municiple Engineers of the Institution of Civil Engineers

Cover photograph courtesy of the Building Research Establishment Contaminated Land Service

Contents

Summary

Construction waste — why minimize?

The construction industry is a major generator of waste, generating more waste than the household sector. Construction waste can account for more than 50% of the waste deposited in a typical landfill.

The cost of waste disposal is increasing rapidly as more stringent controls and taxation are placed on landfill sites. From 1990 to 2000 the cost is expected to have risen fourfold. The disposal of construction waste is becoming a major cost in construction projects. To be competitive, ways of minimizing construction waste need to be

Recycling construction waste — Brunel shows the way! Houses built by Brunel in Swindon from stone arising from the construction of Box Tunnel

Waste — your role and responsiblities

Managing and minimizing construction waste

Who needs to know? Anyone practising in construction, including

- clients, developers, designers, architects, builders, contractors, consultants, civil engineers, industrialists, town planners.

What must they know? Waste needs to go somewhere.

- Preplanning and careful design can avoid waste.

Why? Landfill capacity is a scarce resource and expensive.

- Minimizing waste reduces costs and impact on the environment.

- Criminal liability follows breach of the Duty of Care for waste management.

found. These ways exist — throughout the UK there are a growing number of examples of construction projects where waste has been minimized or eliminated, to the benefit of both the environment and the client.

- Prevent waste by proper maintenance.

- Design with whole-life cost in mind to minimize waste.

- Specify and use reclaimed or waste materials in construction.

- Use techniques which avoid creating waste.

- Reuse waste on site for other purposes or find profitable uses off site.

- Dispose of inert wastes on site.

Some techniques involve the very latest technology, others involve practices that have been carried out for centuries. In all instances, however, advance planning is required.

There needs to be a commitment by client, consultant and contractor to minimize waste.

Construction waste — the law

Waste management is now a carefully controlled and regulated process. These controls have been introduced in response to environmental damage and significant costs that have been imposed upon communities by the illegal disposal of waste.

People handling waste need to know the law and abide by it.

- There are legal definitions of what is waste — interpretation can sometimes be difficult .

- Licences may be needed for the storage of waste.

- Planning permission may be needed for temporary storage of waste, operation of recycling plant and final disposal.

- Carriers of waste must be registered.

- A Duty of Care under criminal law obliges producers of waste to ensure its safe disposal or treatment.

Anyone uncertain of their position should contact the officers of their local waste regulation authority, who will be ready to give advice and guidance.

Waste — the measure of the problem

Over 2 billion tonnes of waste are generated in the European Community each year, of which approximately 500 million tonnes are produced in the UK. The UK waste management industry employs around 100 000 people, and has a current turnover of £5 billion per annum.

Household waste recycling has attracted much attention in recent years, yet construction waste, which represents a greater proportion of the total of all waste, and arguably gives rise to greater problems such as illegal dumping, has not attracted the same sort of consideration, at least not by professionals in the construction industry.

Around half of the waste that goes to landfill can be from construction; a major construction project can completely overwhelm local

Landfilled waste composition in a typical county

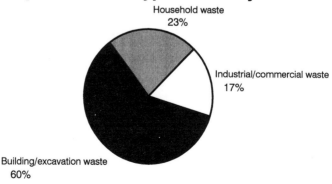

Household waste
23%

Industrial/commercial waste
17%

Building/excavation waste
60%

landfill. Whereas, in the past, responsible professionals would strive to minimize waste generation, now there are also economic pressures; the cost of landfill has doubled over the past five years, and is expected to double again over the next five.

Why is this?

- **Existing landfill sites are filling up.** We have had a legacy of cheap landfill sites handed to us from a time when environmental issues were less prominent.

- **New sites are difficult to establish.** There are often strong objections from local communities to proposals for new landfill facilities. No-one wants a landfill site at the bottom of their garden. However, the fact remains that each household in the UK generates about one tonne of waste per year, and that waste must go somewhere. The construction industry multiplies that need.

A modern landfill site — professionally engineered, scientifically managed, but at an ever increasing cost

- **Higher technical standards for waste disposal are being enforced.** Expensive control systems are now required to be built into landfill sites to prevent the pollution of ground- and surface waters from leachate and to control methane emissions to prevent migration off site, for example into adjacent buildings where a risk of explosion could arise.

- **After-care requirements have become very burdensome.** At one time a landfill owner, once the site was full, could walk away from the site and from any responsibility. Now the site owner remains responsible until the waste regulation authority issues a certificate of completion. This could be 40 years after landfilling has ceased, during which time monitoring and pollution control must continue.

Transporting waste is costly and adds to traffic. The environmental impact of lorries taking material off site and transporting new material on site can be a cause of major annoyance and local opposition.

A modern quarry — by recycling construction materials the need for new quarries and the resultant environmental impact can be reduced

The import and export of construction materials now forms an important part of the formal environmental impact assessment carried out on highway schemes.

Construction and demolition waste can help to meet some of the demand for aggregates. Environmental protection groups have expressed concern over the expansion of quarries and gravel pits in the UK.

By spending time and money in putting construction waste to a positive use we can expect a positive return

- reduced waste transport costs

- reduced waste disposal costs

- reduced expenditure on new materials.

This points to a better environment and increased profits.

There are simple ways of reducing the amount of construction waste produced, and in the UK there are a growing number of examples where firms are tackling the problem. Thought and planning in the construction process is essential. If the same ingenuity and imagination that has been applied to recycling household waste were applied to construction waste, the environmental benefits and the costs saved would be immense.

The statistics show that, by not tackling the problem of construction waste and its reuse, the construction industry has to defend the charge that they are among the biggest abusers of the environment.

What is waste?

In common parlance, waste is a product or material that is un-wanted. In terms of practical waste management, waste is a legal term defined in European and UK law. Anybody who is going to be involved with construction waste must start with the definition 'What is waste', when does a material become waste and when does it cease to be waste. The most important definition of waste comes from the EC Framework Directive which implies that

waste is any material where the holder has an intention to discard the material as no longer part of the normal commercial cycle or chain of utility.

There are many qualifications to the above definition which are subject to interpretation. Detailed below is an explanation of how the definition of waste evolved. The reader can expect to obtain an understanding of the definitions, but because of the legal complexity it is strongly recommended that the advice of your local waste regulation authority should be obtained on specific examples of waste.

The Environmental Protection Act 1990 (EPA 90) Section 75 defines waste as

- any substance which constitutes a scrap material or an effluent, or other unwanted surplus substance arising from the application of any process; and

- any substance or article which requires to be disposed of as being broken, worn out, contaminated or otherwise spoiled, but does not include a substance which is an explosive within the meaning of the Explosives Act 1875; or

- anything which is discarded or otherwise dealt with as if it were waste, shall be presumed to be waste unless the contrary is proved.

The interpretation of the law has been that, irrespective of whether the person who is carrying or holding the material thinks it is of value, a material is waste if it is unwanted by the producer.

On 18 March 1991 the Council of the European Communities adopted Directive 91/156/EEC, amending Directive 75/442/EEC on waste. The new directive, among other things, defined waste, its producer, holder, management, recovery and collection, all of which was to apply in Member countries from 1 April 1992.

Hence, for a time in Britain we had two differing definitions of waste. European legislation usually sets out the broad framework of legislation and it is then up to each of the Member states to take the Framework Directive and transpose it into national legislation in more detail and to accord with common practice in the State. This was the subject of much consultation and discussion. On 24 November 1993 the Minister for the Environment and Countryside announced the Government's intention to implement the Waste Management Provisions of Part II of the Environmental Protection Act.

On 3 December 1993 the Department of Environment issued a paper defining waste for the purpose of waste management licensing. In effect, the Department accepted the main thrust of the Framework Directive and views that its purpose can be summarized as ensuring that substances or objects which fall out of the commercial cycle or out of the chain of utility are treated as waste, and that their collection, transport, storage, recovery and disposal must be properly supervised.

Construction and demolition waste are the most prolific sources of fly-tipped waste, so any exemptions in a definition of waste must be safely distinguished to stop uncontrolled fly-tipping.

The Framework Directive defines waste as any substance or object falling under the categories of waste listed in the directive which the holder discards or intends or is required to discard. The final category of waste listed under the directive 'Q16, Any materials, substances or products which are not contained in the above categories' is a catch-all with the effect that any substance which the holder discards or intends or is required to discard is waste. To avoid any materials becoming waste and therefore coming under the waste regulations and the requirements of licensing, the materials should be reused in a preplanned manner.

According to the Department of the Environment, the main construction items which may be exempted from waste management licensing should be

(a) keeping (storing) waste where it is produced, pending its collection

(b) gathering together and temporarily storing small quantities of waste in connection with its collection for disposal elsewhere, not necessarily at the place of production

(c) concrete, brick and tile crushing, where this is regulated by local authority air pollution control

(d) the deposit of wastes including soil and rock in connection with land reclamation

(e) the manufacture of certain building materials (including roadstone and aggregate), soil or soil substitutes from certain wastes, including construction and demolition waste

(f) storing waste in connection with these activities

(g) the secure storage of moderate quantities of certain waste articles, for example architectural salvage, if they are destined for recovery

(h) storing construction, demolition or excavation waste where it is to be used for construction, either after work has begun or for three months before it starts (repeating the effect of the existing exemptions under the 1988 regulations)

(i) storing road planings at any site provided they are to be used for construction.

Many of these exemptions will be subject to detailed restrictions, especially as to type, quantity and place. It is likely, for practical purposes, that exemptions will be permitted for land reclamation. See

Council Directive 75/442/EEC on Waste — Waste Framework

As amended by 91/156/EEC, with effect from 1 April 1993.

Article 1

For the purposes of this directive

(A)'waste' shall mean: any substance or object in the categories set out in Annex I which the holder discards or intends or is required to discard.

Annex I

Categories of waste

Q1, Production or consumption residues not otherwise specified below

Q2, Off-specification products

Q3, Products whose date for appropriate use has expired

Q4, Materials spilled, lost or having undergone other mishap, including any materials, equipment, etc., contaminated as a result of the mishap

Q5, Materials contaminated or soiled as a result of planned actions (e.g. residues from cleaning operations, packing materials, containers, etc.)

Q6, Unusable parts (e.g. reject batteries, exhausted catalysts, etc.)

Q7, Substances which no longer perform satisfactorily (e.g. contaminated acids, contaminated solvents, exhausted tempering salts, etc.)

Q8, Residues of industrial processes (e.g. slags, still bottoms, etc.)

Q9, Residue from pollution abatement processes (e.g. scrubber sludges, baghouse dusts, spent filters, etc.)

Q10, Machining/finishing residues (e.g. lathe turnings, mill scales, etc.)

Q11, Residues from raw materials extraction and processing (e.g. mining residues, oil field slops, etc.)

Q12, Adulterated materials (e.g. oils contaminated with PCBs, etc.)

Q13, Any materials, substances or products whose use has been banned by law

Q14, Products for which the holder has no further use (e.g. agricultural household, office, commercial and shop discards, etc.)

Q15, Contaminated materials, substances or products resulting from remedial action with respect to land

Q16, Any materials, substances or products which are not contained in the above categories

Is it waste?

Is the material, substance or object:
● a threat to human health or the environment?
● being discarded?
● no longer part of the commercial cycle or chain of utility?

Has the client, designer, contractor, subcontractor or resident engineer failed to preplan the further use of the material?

❮ **No to all questions**

Yes to any question ❯

Unlikely to be directive waste

Directive waste →

Any doubts – contact the local waste regulation authority

the diagram above for other exemptions. In all cases any queries should be referred to the relevant waste regulation authority.

When is a waste management licence required?

A waste management licence will be required for

(a) landfill; that is, the disposal of waste by permanent deposit in or on land

(b) keeping waste that is not going for disposal; except for small-scale exemptions, a licence will be required for keeping waste such as spoil

Exemptions

Exemptions from controls may be permitted for land reclamation provided the substance or object is being used for genuine beneficial use as against relieving the holder of the burden of disposal. Would the user seek to use a substitute if the material was not available?
Exemption is possible for construction waste provided it is spread on industrial land or another development which is incapable of beneficial use without such treatment, is spread in accord with a separate planning application and with a quantity of not more than 20 000 m^3 per hectare.
Other exemptions are for holding bricks, blocks, roadstone or aggregate for treatment at the place produced or where used,

if it is not at the place where the waste is produced and is destined for disposal and not reuse

(c) keeping waste off the site where it is to be used; with special exceptions, such as architectural salvage and road planings, storing any construction, demolition and excavation waste away from both the site where it originates and the site where it is to be used will be licensable

(d) special waste; none of the exemptions apply to special waste (see Glossary, Appendix 1)

(e) activities beyond the bounds of an exemption; where an otherwise exempt activity takes place beyond the prescribed scope of an exemption, for example beyond the limits the exemption sets on the type or quantity of waste or the time, then all that activity is licensable.

Is it waste?

It is not likely to be 'waste' if the material has been preplanned for reuse, and the holder (client, developer or contractor) can show it is to be used beneficially and without danger to health or the environment. Otherwise, it will be regarded as 'waste', will need a waste management licence and be required to be taken by a registered carrier to a licensed site.

If you want to avoid the need for waste management licences and waste regulation, be sure to preplan the gainful reuse of any surplus materials and to minimize the waste you create.

Waste minimization

Objectives

If the amount of construction waste going to landfill can be reduced

- environmental damage from quarrying can be cut

- energy consumption on transport can be reduced

- landfill capacity can be conserved

- profits on construction can be increased through reduced disposal costs.

Waste minimization

Directive 91/156 EEC

The law on waste in the UK is founded on European directives. In particular, Directive 91/156 EEC sets out the hierarchy of waste management. First, the prevention or reduction of waste by

- the development of clean technologies

- the improved development and marketing of products designed to produce little or no waste.

Second, the recovery of waste by means of

- recycling, reuse or reclamation and the use of waste as a source of energy.

The UK Government is required to take measures to prevent the endangering of human health or the use of processes or methods that could harm the environment. Measures to prohibit the abandonment, dumping or uncontrolled disposal of waste must be taken.

The waste hierarchy applied to the civil engineering field becomes

- **Reduction.** Designing-out waste, balancing cut and fill.

- **Reuse.** Shuttering.

- **Recovery.** Recycling, for example bitumen macadam planings, broken and crushed concrete, cleaned recovered bricks, composting, energy.

- **Disposal.**

Some examples

- **Reduction/design.** Material wastage, possibly from temporary works or over-design, is simple to avoid if considered at conception stage and integrated as an essential part of professional design. An overlying commitment to consider and thence reduce waste is the foundation to tackling the issue. Cement and lime stabilization can be used to improve the load bearing performance of certain soils which otherwise would need to be excavated and cleared off site.

- **Reuse.** Directly re-employing materials after refurbishment or in lower- grade applications is simple and effective. Unwanted street furniture can be sold to specialist architectural salvage companies rather than scrapped. Excavated soil used for landscaping or noise bunding is an easy option for road builders.

- **Recycling.** Concrete crushed on site to produce secondary aggregates reduces transport of bulky, low-value goods. Mobile, temporary plant is readily available, but needs to have space allowed for it at the earliest stages of most contracts. Scrap metal is an established reprocessing operation, and reinforcing bars are suitable. Self-employed registered carriers who deal in scrap materials can help where the market is weak.

- **Dispose of beneficially.** Brick rubble can be used to harden farm gateways, but can be open to misinterpretation. Take care that your idea of beneficial coincides with that of other interested parties. Broken-up bituminous material from roads can improve the surface of footpaths and bridleways if wisely used.

- **Recover energy.** Construction waste is mainly inert, and the question of energy recovery rarely arises. There may be an opportunity to provide unwanted brushwood, etc., to outsiders as fuel for heating in rural areas. The manufacture of many construction products such as bricks and cement involves the use of much energy; recycling these materials recycles the energy. Even in transporting waste soils there are significant energy costs, for example shifting 20 tonnes of soil 10 km will cost well over £2 in energy used.

- **Dispose of to landfill.** If all else fails then landfill is the only option left. Even here it can be used as a means of restoring industrial scars, but really only gives marginal benefits when compared with the options listed above.

Economics and transport costs

Is recycling and minimizing construction waste financially viable? The financial benefits include

- reduced costs for the transport and disposal of waste materials

- reduced costs of using new materials

- increased returns from selling waste materials for reuse.

High-value materials such as copper and lead which are valued in thousands of pounds per tonne are obvious candidates for recycling. However, the value of these materials can fluctuate widely. For lower-value materials, haulage costs become significant in the over-

Managing construction waste — alternative approaches

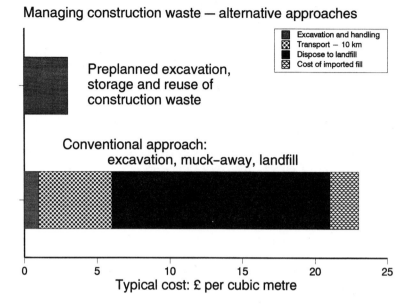

Typical cost: £ per cubic metre

all financial balance. If you cannot afford the cost of transporting the material very far, then the potential market is that much smaller. Because of its high value, copper can be sold into a national market; however, lower-value materials such as waste soils may only justify transport over a few miles. Low-value materials therefore require local customers.

The value in recycling low-cost materials such as soils and secondary aggregates is the saving of the cost of disposal, and, if the materials can be reused on site, the minimizing of haulage and handling costs. This saving can be substantial — typical costs of landfill have doubled in the last five years, and may double again in the next five.

There are several types of costs and benefits, the first being financial and non-financial. There are many things in life which are formally valued in cash — goods and services, salary, etc. However, many of the most important aspects are not formally valued, for example friendship, pleasure, happiness, or anxiety, uncertainty,

	Costs	Benefits
Client	£ New materials £ Transport £ Waste disposal	Minimization of construction waste leads to: £ - lower costs and £ - INCREASED PROFIT
Community and environment	£ Fly-tipping clearance ○ Contamination of land by illegal disposal ○ Construction traffic nuisance ○ Construction noise ○ Development of new landfill sites ○ Expansion of quarrying ○ Visual intrusion ○ Community stress ○ Loss of habitats	○ Landscape enhancement

£ Financial costs
○ Non-financial costs

security, safety, fear, as well as pollution, landscape intrusion, noise, etc.

Economists draw a further distinction between internal and external costs and benefits. Internal costs and benefits are the costs and benefits exchanged between consenting parties; money often passes between the parties involved. External costs and benefits are the side-effects of an activity, where other parties are affected either beneficially or adversely: they receive a benefit without paying for it, or suffer a cost without being compensated for it. Examples include environmental or social costs, such as noise and visual intrusion, or the landscape benefit to nearby residents of restoring adjacent derelict land, or providing a noise-reducing bund. Obtaining a value for external costs and benefits is very difficult, but this does not mean that they are insignificant. Recent years have shown that people do care greatly about these externalities, sometimes enough to risk a jail sentence.

The increasing environmental awareness of the public has focused attention on the importance of external costs. As time passes, increasing waste disposal costs will require the minimizing and reuse of construction wastes to become an integral part of the construction process. Even when it may not be financially viable to reuse and recycle, when the external costs and benefits are taken into consideration the balance may shift the other way. Dealing sensitively with waste problems can help ease the passage of a construction project through the planning process, and reduce concerns of local residents. With the environmental assessment regulations it is a compulsory part of the design process.

Waste and the law

The Duty of Care and the Environmental Protection Act 1990

Registration of carriers under the Control of Pollution (Amendment) Act 1989

Any carrier of controlled waste is required to be registered with the relevant waste regulation authority (excepting waste collected by charities and voluntary organizations). The waste regulation authorities now hold the registers of carriers; that is, all carriers who are legally entitled to carry controlled waste. The carrier of a controlled waste may be stopped to have the waste inspected, and be required to produce the certificate of registration within seven days (see COP(A)A 1989). If an offence is committed under these regulations the carrier may be fined up to £5000 for each offence. The waste regulation authorities have the ultimate sanction of applying to a magistrates' court for the seizure of the vehicle involved in fly-tipping. The regulations have been in force since 1 April 1992.

Duty of Care

Holders of controlled waste must

- be able to describe the waste being held

- keep the waste under proper conditions of containment

- ensure the waste is transferred to an authorized person for authorized transport purposes;'authorized' being a registered carrier

- issue with the waste being transferred a waste transfer note that clearly describes the waste.

The transfer note must contain the following details

- identity of waste

- quantity of waste

- containment, for example whether loose or in a container, and, if the latter, the type of container

- the time and place of transfer

- the name and address of the transferor (the person who is dispensing with the waste)

- the name and address of the transferee (the recipient of the waste)

- whether the transfer is to a person for authorized transport purposes.

Recipients of waste must

- ensure that the waste transfer note is properly completed before accepting the waste

- ensure the waste is properly contained prior to further handling, reclamation or final disposal.

A transfer note is required every time waste is transferred from one person to another. There may, however, be special arrangements for identical repetitive loads.

The Duty of Care is embodied in the Environmental Protection Act 1990. Under Section 34 it shall be the duty of any person who imports, produces, carries, keeps, treats or disposes of controlled

Shortly after the introduction of the registration of carriers, a lorry driver under contract to a County Council approached a householder to ask if he knew of a local farmer who would be prepared to take a load of road planings. The householder was in fact the County Surveyor and also the officer in charge of waste regulation. The lorry driver was 'unlucky'.

waste, or as a broker has control of such waste, to take all such measures applicable to him in that capacity as are reasonable in the circumstances

- to prevent any contravention by any other person of Section 33 of the EPA 1990

- to prevent the escape of the waste from his or her control or that of any other person

- on the transfer of the waste to secure that the transfer is only to an authorized person, or to a person for authorized transport purposes, and that there is transferred such a written description of the waste as will enable other persons to avoid a contravention of that section and to comply with the duty under this subsection, as respects the escape of waste.

The code of practice for the Duty of Care has statutory standing, although admissible and liable to be taken account of in court, it is subordinate to and cannot change the meaning of legislation contained in Section 30 of EPA 90 and the 1991 regulations on Duty of Care contained in Statutory Instrument 2839. An offence may be committed when reasonable steps are not taken, whether or not illegal disposal of the wastes in question subsequently occur. Adherence to, or breach of, the code may be used as evidence as to the criminal liability of a person transferring waste, but the breach of the code is not of itself an offence. An offence is committed when you break the law as laid down in the Acts and Statutory Instruments. The law, as well as requiring the use of registered carriers, requires that you act reasonably in your handling of waste. The code of practice for the Duty of Care on Waste is the official guide to what is reasonable. On this you will be judged!

In simple terms

Your waste must be taken by a registered carrier and disposed of properly at a licensed site.

Until the introduction of the Duty of Care and Registration of Carriers the apocryphal source for broken concrete in the London area was 'work going on at Heathrow Airport'. The airport would have needed to have been rebuilt many times to produce this quantity of gold-plated secondary material.

Civil liability

All engineers should keep in touch with developments on their legal responsibilities. It is a rapidly developing area, and will have a direct financial effect, at the very least in terms of the cost of professional indemnity premiums.

Only in recent years have environmental controls been sufficient to act as a reasonable safeguard against pollution. Unfortunately, Britain has a legacy of centuries of uncontrolled industrial activity, such as gas works, oil refineries, scrap metal yards, chemical works, mining and so on. Typical contaminants that can be found are asbestos, brickwork impregnated with chemicals, oil, tars, phenols and heavy metals. Certain industries, such as luminous watch making and fabrication of gas mantles, are associated with radioactive materials. Demolition waste from these sites may occasionally be contaminated and will need proper disposal.

The European Union is currently consulting on a directive on Liability for Environmental Damage. There are three main definitions of liability used

- Absolute liability, where a party is absolutely liable for the conduct of an asset or action irrespective of act of God.

- Strict liability, where a party is liable without fault being proven against him, and (generally) that the damage or loss was foreseeable in the legal meaning of that term.

- Fault liability, the plaintiff has to prove on the balance of probabilities that the party is liable.

Insurance cover for environmental liability is likely to be a problem if strict liability is passed.

Definitions of environmental damage will be developed during the consultations. Pollution of the environment in the EPA 1990 is defined as

capable of harm to man or any other living organism supported by the environment. Harm means harm to the health of living organisms or interference with the ecological systems of which they form part. And in the case of man includes offence to any of his senses or harm to his property.

Waste and the planning system

An outline of the planning system

There are differences in the waste planning system between metropolitan and London boroughs and shire counties, as is indicated in the diagram. In the former, waste planning is contained in a unitary development plan, while in the shire counties the policies are contained in a structure plan, with the details in a waste local plan.

National planning policies, regional guidance and the provision of structure plans set the broad framework for planning at the local level. Local planning authorities have a duty to prepare structure plans or unitary plans with broad policies on waste. Local plans set out detailed policies and specific proposals for the development and use of land and should guide most day-to-day planning decisions.

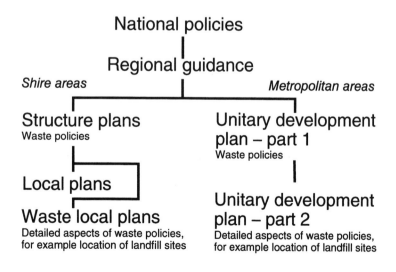

National policies

Regional guidance

Shire areas   *Metropolitan areas*

Structure plans
Waste policies

Unitary development plan – part 1
Waste policies

Local plans

Waste local plans
Detailed aspects of waste policies, for example location of landfill sites

Unitary development plan – part 2
Detailed aspects of waste policies, for example location of landfill sites

Case study — reuse of material excavated from a motorway

During the 1980s, a series of contracts were let by Buckinghamshire County Council as agents for the Department of Transport to reconstruct and strengthen the M1 motorway. Engineers working for the County Council thought it would be advantageous to lease a piece of land near Newport Pagnell, obtain planning permission and allow contractors to use the area to tip concrete and bituminous-coated material.

A large number of contracts were being let for new road construction in the vicinity by the Milton Keynes Development Corporation, in which the County Council had a financial interest, where the roads would become adopted highways. The County Council also planned to build Fenny Stratford Bypass, for which the material could be used as a capping layer for embankments.

In the event, 100 000 m^3 of material were delivered, stored and crushed at the site and reused on neighbouring roadworks. After payment for the land and crushing of the excavated material, there was a significant financial benefit to the County Council.

Objections were received from a number of public organizations and authorities on environmental grounds. Such objections need to be satisfied if such a scheme with its environmental advantages can proceed. The engineer has to be aware early on of the possibilities of financial and environmental benefits and be motivated professionally. Also, elected members have to trust the professionalism of engineers and be prepared to give political backing to such a scheme for a sustainable development.

Unitary development plans (UDPs) have two parts. Part 1 is similar to the structure plan in a non-metropolitan area and contains a written statement of the authority's general policies for the development and use of land in their area. Part 1 provides the framework for detailed proposals in Part 2, which is similar to the local plan in non-metropolitan areas.

Minerals local plans outline the forward policies which provide for the supply of minerals and ensure the required degree of environmental protection associated with their development. Waste local plans (WLPs) appear in the UDP in metropolitan areas. In all other areas of England, counties may either prepare a waste local plan or combine it with the minerals local plan. The waste local plans should include forecasts of waste arising and estimates of movements and flows of waste into and out of the authority, the identification of physical planning constraints

The problem: waste spoil and construction waste on site

Reuse and recycling
If the material can be used as a raw building material it may still be waste, so check with the WRA
Materials can be taken to a site for construction purposes provided the use of the material is planned in advance and included in the specifications of the construction works.

For further reading see DoE Circular 11/94

For further advice and peace of mind, contact the local waste regulation authority

Concrete crushers
Contact the WRA and the planning authority to check on the latest position regarding licensing and planning exemptions

Temporary storage on site
Where waste is being stored for planned and specified use at a later stage in the construction project, exemptions may apply and licensing may be unnecessary

Landfill
Ask to see the site licence, and check that it covers the type of waste you are depositing — many sites will take inert construction waste but may limit the organic content, for example paper, wood and plastic, some may not take oily wastes, tyres or calcareous materials. Few sites will take special waste such as asbestos or chemicals

Special dedicated site
- Landscaping • Reclamation
- Noise bunds • Land raising
- Land forming

Check with the WRA whether you need a licence. Licence applications can take 3—6 months
Obtain planning permission. This can take 3—6 months or longer

It is illegal to start operations before the necessary licensing and permissions have been obtained

on the location of waste disposal and waste management facilities, and identify potential sites for the disposal of waste. WLPs should have reference to waste disposal plans for their area and be in conformity with the structure plan.

Section 50 of the Environmental Protection Act 1990 obliges waste regulation authorities (WRAs) to prepare waste disposal plans (WDPs); this section came into force on 31 May 1991. The DoE wished to receive draft WDPs by the end of 1994. Planning authorities will use WDPs as an information source to prepare waste local plans or, in the case of metropolitan authorities, the waste disposal aspects of the UDPs. WDPs must set out the arrangements made, and proposed to be made, by waste disposal contractors for the treatment or disposal of controlled waste. WRAs are to classify wastes in accordance with the European Waste Catalogue (EWC).

Waste and the planning system — practical advice

If the material you are taking to a site can be used as raw building material then it may still be classified as waste, so take care to check with the relevant WRA. Even if it clearly is waste, it is still permissible for it to be used for construction purposes if the specification permits it. If this is the case, then waste can be taken to a site for construction purposes. (Make sure, however, that the construction itself is permitted. If it is used for something that is lacking planning permission you are still liable to prosecution for illegal tipping.) Beware of farmers' hardstandings which have maximum permissible area, but which tend to be over-generously filled. If you don't get written instructions to tip somewhere then you will be assumed to be the one at fault, and the fines are punitive.

The other option is to take your waste to an already licensed site. As far as construction is concerned, there are generally three types of operation

- Recycling areas, where waste is separated out to form new materials.

- Transfer stations, where waste is bulked up and moved by more economical means.

- Final disposal sites, which generally speaking entails landfilling of the material.

There are minimum requirements for the conduct of all sites, and particularly the types of material that can be accepted by them. You will be required to detail the nature of the wastes you intend to tip and are required by law to tip only that which you have described.

When considering using a site you should ask to see a copy of the site's waste management licence. If they can't produce one then you should draw your own conclusions as to the likelihood of legal compliance being achieved. On the licence they produce check the types of waste that may be accepted and make sure your waste will fit within the parameters laid down.

Many site waste management licences will not permit the deposit of oily wastes or tyres. Few will accept special waste such as fibrous asbestos or chemicals. Many will allow construction spoil, but may limit the 'non-mineral' content such as paper, plastic and wood. Some may not permit calcareous materials. These restrictions reflect the environmental conditions pertaining to the site, so unless you want to cause pollution and wish to risk being prosecuted, you should keep to the terms specified in the licence. Once selected, keep a copy of the licence, or at least the licence number, on your files for future reference.

Client's checklist

Much spoil and waste from construction goes to landfill unnecessarily at a cost to you and a waste of a precious resource. Some 50% of landfill is at present used up with spoil and construction waste. It need not be so. In order to build your construction at the lowest overall cost you should brief your designer and contractor so they know that you wish them to design out the problem of waste and its disposal cost.

- Waste prevention and minimization should be part of the brief — ensure the brief requires the designer to consider early in the design

Client takes the lead! — the RAC Centre near Bristol is an example of a building where the client wished waste to be minimized. The designer eliminated the waste arising from the project by raising the level of the car park.

---

process the creation and after-use of spoil and construction waste, and that he and any contractors comply with the Duty of Care.

- Use adjacent land for landscaping — ensure the brief identifies any adjacent land which you hold which can be landscaped using spoil or otherwise waste material, or on which materials can be stored for reuse on or off the site.

- Marketing advantage — consider the marketing advantage that may be created by minimizing construction waste.

- Beware contaminated land — if land is contaminated by previous processes consider whether the development can be built safely by leaving the material in place subject to appropriate remediation. Contact the local waste regulation authority for advice or the local planning authority for information about contaminated land.

- Use a specialist demolition contractor — if demolition is involved ensure a specialist contractor is used to carry it out safely and with maximum reuse, recycling or sale of the materials.

- Avoid over-specification — specify carefully and look to the likely long-term use of the construction so that the designer only designs for those uses and does not over-specify the construction.

- Be aware of the Duty of Care — ensure that you are aware of the Duty of Care under the Environmental Protection Act 1990 and any spoil or waste is taken from a site by a registered carrier for legal reuse or disposal by the holder of a waste management licence.

- Determine the details of the proposals for reuse and/or disposal and check the details with the waste regulation authority where there is any doubt about any part of the arrangements.

- Avoid commissioning work which will lead to earthworks in the wet winter months.

Creating a market for construction waste

Who creates markets for products? Generally, this is achieved by either the client specifying a new item, or else through a manufacturer persuading designers that their new idea is a good one. At present there is a dearth of either sort of initiative.

There is an unfortunate tendency towards slavish copying of previous specifications in the construction industry. This is normally the result of a lack of understanding of the needs of the particular task, and leads to over-conservative design. With a better understanding of the project it is often possible to dispense with some traditional items that have been left in the specification long after they have become redundant. Better understanding also leads to the confidence to specify the right material for the job, not just one that is more than good enough. Over-specification leads to waste just as surely as over-ordering of materials does. The difference is that over-ordering leads to unwanted heaps, is easily spotted, rectified and subsequently avoided. Over-specification, on the other hand, is hidden by the excess being used up in a wasteful manner and often goes undetected.

Designer's checklist

With up to 50% of all landfill being used for spoil and construction waste and the proper requirements of the Environmental Protection Act 1990 increasing the cost of landfill, designers have a duty to their client and the public to minimize construction waste and so save costs and the misuse of a valuable resource.

- Review the client's checklist.

- Construction waste is a resource — be aware at all stages of design and construction that 50% of landfill can be comprised of spoil and waste from construction.

It can be done! — a building in Copenhagen made from recycled materials

- Transporting and disposing of waste is expensive — be aware of the local cost of transport and deposit of waste from construction. Many informed sources foresee the cost of the latter doubling in the near future.

- Find landscape uses for waste soils early on — consider with the landscaper at the start of the design how otherwise unsuitable material could be used, so saving the need to take it off site. Time is needed to permit the acquisition of the land required for this purpose.

- Minimize construction waste by
  - ○ Not excavating spoil material or concrete or other materials which can be left in place.
  - ○ Attempting to recycle as part of the design, for example by providing sites and opportunities for crushing concrete, etc.
  - ○ Designing-out the problem, for example by building a mass earth dam against the face of a leaking concrete dam, saving demolition and deposit of the old concrete.
  - ○ Balancing cut and fill.
  - ○ Using material which would normally be unsuitable for construction in areas where material strength is not required, for example filling in the centres of roundabouts or noise bunds, or use lime/cement stabilization to increase load bearing. Avoid earthworks in the winter months when the ground is liable to be waterlogged.
  - ○ Ensuring contaminated material is safely dealt with or avoided by leaving in place if it is safely dealt with, for example by raising levels of the construction or providing a pelican crossing instead of a subway that goes through contaminated soil. Driven piles may leave contaminated material in place while bored piles may produce contaminated spoil which requires special disposal.
  - ○ Using contractors or subcontractors who can and will reuse materials on site or elsewhere, for example use a demolition contractor who is a member of the National Federation of Demolition Contractors.
  - ○ Specifying materials to the performance required, for example permitting the use of broken bricks as a capping layer to a road embankment.

- Design and supervise the construction constructively not dogmatically.

- BS7750 — consider whether your organization should achieve and demonstrate sound environmental performance as defined by BS7750, *Specification for Environmental Management Systems.*

- Follow the Duty of Care — ensure that you are aware of the Duty of Care under the Environmental Protection Act 1990 and any spoil or waste is taken from your site by a registered carrier for legal reuse or disposal by the holder of a waste management licence.

Contractor's checklist

- Review the client's and designer's checklists.

- Adopt a policy for minimizing waste — consider registering for BS7750 or adopting an environmental charter.

- Bidding — indicate to the client what savings might be made on the contract if additional work is authorized for waste minimization.

- PR benefit — note the benefit from using recycled construction materials and being described as green contractors/consultants.

- Ensure that planning and site staff are trained in the requirements of waste regulation and in the advantages of waste minimization.

- Identify likely wastes.

- Identify local markets/uses for wastes or treatments that would turn the waste into a marketable or useful product. The local waste regulation authority may have a list of firms who might help.

- Identify customers as early as possible and obtain legal and financial agreement for them to take the waste before work starts.

- Check and assess the costs of disposing of waste and the alternative disposal options — identify potential sites, transport costs and likely environmental impact.

- Identify storage/process areas required for waste and apply for necessary planning consents and waste management licences.

- Understand the chart of licence application/planning approval process for temporary storage/mobile crusher plants, etc. (see figure on page 28).

- Check your proposals with the local waste regulation authority for compliance with law and policy.

Practical waste management and minimization — minimizing waste from maintenance

General points

- Keep ahead on preventive maintenance to forestall more serious repairs.

- Design-out problems before you arrive on site.

- Use a maintenance method that minimizes waste generation.

- Find positive uses for any waste materials generated.

- During works, avoid contaminating material proposed for recycling/reuse.

Sewer repair

- Design for maintenance, not just for construction.

- Use 'no dig' technologies to refurbish pipework without excavation.

- Employ regular closed circuit television survey of defective parts of a sewer network to determine the right time for refurbishment. Can repair be delayed until the water-table is at its lowest so that excavations produce less wet and unusable back-fill?

- During excavation keep the wastes separate. Keep

○ the surfacing for use as rubble
○ dry backfill for the top of the trench
○ the wet backfill for the bottom (if appropriate)
○ the pipe, bed and surround to go as waste off site (therefore easier to confirm it has gone off site and not gone back as unsuitable backfill).

- Avoid dumping spoil on topsoiled verges. Strip them first and reduce imported topsoil and stone picking.

- Direct excess materials to storage areas, not to landfill. Keep the materials separate.

Bitumen macadam roads maintenance

Scarcity of funds for road maintenance has caused highway engineers to look for lower-cost methods to provide a surface which can provide most of the requirements of high speed and heavy traffic; that is, a waterproof, non-skid surface. Somewhat fortuitously this has resulted in the reuse of much of the existing material, which reduces the amount taken to landfill.

Poor skid resistance may be solved by

- Burning off the fatty top surface.

- Scabbling to provide a new running surface.

- Surface dressing, where carefully selected stone chippings are laid on a thin layer of bitumen or polymer emulsion and then rolled in to become an integral part of a professionally designed new road surface which is both waterproof and skid resistant.

- Surface treatments of, for example, bauxite chippings, which can greatly improve the skid resistance on some highly stressed road surfaces.

Water penetration is a major cause of road failure. Avoid the need for major reconstruction and the consequent cost and waste generated by keeping ahead on preventive maintenance.

Hot mix recycling

Road planings are mixed with new materials and hot binder in a drum mixer. Early indications are that the material performs as well as new mixes.

Cold road recycling

- Method 1 — rotovate existing material in situ. Re-rotovate, adding hot bitumen with steam (foam bitumen and additives).

- Method 2 — rotovate existing material in situ and add emulsion and cement.

- Surface dressing — where levels permit, add on top of existing road surface, and avoid scarifying, which itself generates waste.

- In-situ recycling — take off 20 mm of the surface, reprocess and relay.

- Slurry sealing is a low-cost method appropriate for footways.

None of the above methods significantly improve strength. Options to consider are

- Resurface with 40 mm of new material. Too often the old surfacing is excavated in order to lay the new one, with the old material being taken to landfill which increases construction waste and loses the residual strength of the old material. Professional highway engineers should always, if possible, lay new surfacing on top of old.

- In-situ recycling. Many successful trials have been carried out with bitumen material recycling on all types of road, from minor housing estate roads to trunk roads and motorways (the A5 and M40 being examples).

- Geogrids have also been a success in ensuring that the existing materials are in the main reused to provide a long-lasting new road or footpath.

Alternative uses can be found for waste materials

- Store until positive use can be found. Where old road materials cannot be reused for in situ recycling, they can be taken off site to be stored until a positive use is found, perhaps in another road repair job, for example as a capping material or as clean fill.

- Use to improve footpaths. Old road materials can be used to provide a more stable surface for rural footpaths and byways, but sensitivity is needed or both landscape and ecology may suffer. The objective is to improve the footpath, not to dispose of waste. Discuss the use of the materials with the local highway authority, who are responsible for both roads and footpaths. Consult where necessary with local environmental groups.

Concrete roads maintenance

Anyone who has had to break up an old concrete road will know how expensive it is in time and money. Many of the maintenance approaches for bitumen roads can be applied just as well to concrete roads.

For excess smoothness, which can cause aquaplaning , use

- surface dressing
- scabbling.

Poor friction can be resolved by

- surface dressing
- bush hammering.

For excess noise

- overlay with blacktop — on high-speed roads tyre noise can be reduced by 3 – 8 dB.

For joint failures

- repair the joint, rather than employ full reconstruction.

For settlement of slab sections

- level with pressure grouting

- overlay with bitmac to obtain a level surface.

The following are techniques which have been used in the case of major base failure, for example where slabs are cracked and sub-grade is 'pumping up' towards the surface, discolouring the concrete.

- Reconstruct — consider using resultant broken concrete to exceed specification for base/formation layers.

- Grahamization — a proprietary technique which involves the hairline fracturing of old concrete slabs to enable them to be left in situ and used as an equivalent sub-base.

- Build over — it is sometimes possible by careful examination of levels to leave much of the old road in place and reconstruct the new road on top of the old.

Off-site uses for waste materials are those such as separating reinforcing bars and recycling as scrap.

Composite road maintenance

Separate the bituminous and concrete factions and recycle accordingly.

Bridges

For the case of bridge understrength, use the existing structure to best effect using techniques such as

- narrowing — reducing lanes/lane width can meet strength criteria as well as usual traffic flows

- load tests to prove the strength of the bridge

- light aggregate concrete decking to reduce dead load

Brick arch strengthened with tie rods

- replacing part of the bridge

- fencing off weak areas, for example verges, where assessment has indicated understrength.

If replacement is required, consider using the existing bridge as shuttering.

For the case of bridge undercapacity the existing structure can be widened, provided there is sufficient strength within the bridge.

Complete reconstruction of masonry arch bridges can often be avoided by providing

- a new concrete invert to strengthen the arch and/or tie rods

- underpinning to piers and abutments.

A strength-and-cost assessment should always be carried out on such a structure to ensure that the public get the best value for money while at the same time reducing construction waste.

Dams

Even dam structures can be reused when they come to the end of their life or need major reconstruction. The Upper Glendevon Dam is a mass concrete structure built to supply the Fife Region in the early 1950s. There were concerns as to its stability, leakage and response to seismic action, which severely restricted its impounding capacity. The option chosen to restore the dam to full impounding capacity was to stabilize the dam by forming a rock fill embankment against the downstream face using 250 000 m^3 of rock quarried 200 metres from the dam.

Composite structures — demolition

Does the structure need to be demolished? Could refurbishment or alternative uses avoid the need for demolition?

Research the structure being demolished and consider the following points

- What was it made of?

- When was it made?

- What stresses are concealed in the structure?

- What has been the history of the structure, were there accidents, faults, repairs, etc.?

- Identify materials with potential for reuse.

Note that even if no records or 'as built' drawings can be found, the designer or resident engineer may still be alive and have photographs, recollections or even calculations. The ICE archives or the Panel for Historical Engineering Works may also have records.

Prepare a demolition plan.

- How to manage the demolition?

- How to manage health and safety?

- How to deal with and market resulting wastes?

During demolition, separate materials for recycling/reuse and keep them clean and uncontaminated. Individual clean materials have their own market.

Practical waste management and minimization — minimizing waste from new construction

General principles

Plan in advance

- obtain consents for storage of waste

- identify markets for waste materials.

Design to minimize waste generated and use recycled materials.

Groundworks

Contaminated land and special waste

If you suspect the land is contaminated, do not disturb it until you have

- Conducted a desk study of past uses.

- Conducted an environmental site investigation in consultation with the local waste regulation and planning authorities.

- Evaluated the options, including avoidance of the site, treatment in situ or disposal to special waste landfill. Consider design options which isolate the contaminants, for example raising the level of a new road to avoid the need for excavation. Special waste must be disposed of at a licensed site. The dumping of special wastes is a criminal offence and there are

Special waste disposal
A simplified worked example

Disposal costs
Conventional waste: £10/tonne
Special waste: £100/tonne

Problem
1 tonne special waste and 100 tonnes conventional waste
Disposed of separately = £1100
1 tonne of special waste and 100 tonnes of conventional waste mixed together = 101 tonnes of special waste
Disposed of together = £10 100

severe penalties for companies and individuals who break these laws, including unlimited fines and jail.

Do not mix special and ordinary waste as mixing even a small amount of special waste with ordinary waste will result in the entire load being classified as special waste. Special waste can only be disposed of at a limited range of licensed sites and the cost of disposal is ten times or more the cost of disposing of conventional waste.

Remember

Special waste + conventional waste = special waste.

Soil from excavation

- Consider using subsoils as topsoil. Low nutrient levels, such as those provided by grass cutting, can encourage wild flowers and reduce maintenance requirements. However, professional advice should be obtained as not all subsoils are suitable.

- Upgrade subsoils to topsoils by mixing with composted municipal garden and green wastes.

Special wastes often encountered in new construction work

- Paint

- Strippers

- Solvents, acetone, turps

- Epoxy resins

- Concrete additives

- Wood preservatives

- Petrol

- Oils and tars

- Phenols

- Asbestos

- Heavy metals

Sewers

Damage to sewers is often the consequence of heavy loads, under-design, inadequate bedding or penetration by tree roots. Pipes should be designed to withstand foreseeable loadings.

- Provide excess capacity where possible to prevent relaying earlier than is avoidable.

- Provide space to place wet materials temporarily. Given time, many will drain out and be reusable if carefully managed. Keep unwanted contamination out and keep the rain off with sheeting. Every cubic metre kept on site is over one cubic metre not dug out and one cubic metre of landfill not consumed.

- Employ a forester instead of a bulldozer to clear the way through woodland. Logs have value, thinnings can make fence

Road widening on the M20

posts and even brushwood can be used if made into excavator mats or chipped for soil conditioning.

Buildings

- Carry out effective desk studies of all past uses of the site. Look as far back as possible; a day spent locating the site history can save weeks of conflict and cost.

- Avoid excessive excavation of foundations, particularly if the site is possibly contaminated. Make the design flexible enough to allow a choice of foundations in the event of contaminated soil being encountered.

New road construction and widening

Designers have a responsibility to design with waste minimization in mind and inform the client of the options.

Waste and development planning stages

New road and road widening proposals should be included in waste local and structure plans, including proposed sites for temporary storage of wastes. If the road improvement is significant enough for inclusion in a structure plan, the waste created should be included in the waste local plan.

Design stage checklist

- Aim to balance cut and fill.
- Minimize creation of unsuitable materials and look for reuse options.
- Evaluate potential for noise barriers and landscape mounds.

- Is it necessary to remove existing mature trees or can they be left unharmed?

New road construction — traditional approach

New road

= Waste soil

Alternatives

Noise reducing bund

Natural contour

Natural contour and noise reducing bund

1 : 6

Grade-off to permit return to agricultural use

Carriageway widening

Same level
- High waste
- Significant visual intrusion

Split carriageways
- Low waste
- Landscaping retained

- Follow contours, consider separating carriageways.
- Consider treatment options for unsuitable materials which would enable their use within the project.
- Where there is an opportunity to use imported waste materials in landscaping mounds, ensure that the necessary waste disposal licences are obtained before commencement of the contract.
- Can gradients be increased to avoid the need for deep cuttings or high embankments?
- Can recycled materials be specified for use in construction.

Construction

Imaginative design should be capable of significantly reducing costs at the construction phase. To do this it may be necessary to obtain planning permission to use variable heights of mounding so

Steeper gradients may reduce or eliminate cuttings and embankments and reduce visual impact

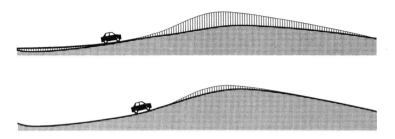

that if more waste arises than originally envisaged it can be incorporated on site (an example might be a greater than expected quantity of unsuitable material).

If there is space to store the material reserved on site then it is likely that the contractor can use it to store material for reuse later in the contract.

Excavated material is often of variable quality for load carrying. It is worthwhile to plan for this by having landscaping areas where variable quantities of material can be used, which would save large quantities being sent off site as construction waste. Such areas can be used for 'unsuitable' material, where it may in suitable circumstances dry out for use later in the contract or be used to widen noise bunds or reshape mounds.

Certainly, a more imaginative approach to planning permission is needed for mobile crushers and temporary stockpiling if there is to be a significant decrease in construction waste. Demolition firms cannot be expected to reuse concrete if they are not allowed to erect crushing plants for the duration of a large contract.

Standing timber and brushwood

Avoid creating waste by reusing on site, for example using chipped brushwood and garden-type waste to produce ground mulch, or preserving existing trees.

Appendix 1.Glossary

Authorized persons (for the purposes of the transfer of waste under the Duty of Care) — a waste collection authority, the holder of a waste management licence (or a person exempt from the requirement to be licensed), a registered waste carrier (under Section 2 of the Control of Pollution (Amendment) Act 1989), or a person exempt from registration.

Biodegradable — material capable of being broken down as a result of the action of micro-organisms.

Building excavation waste — waste arising as a result of construction, demolition and excavation work (including waste arising from improvement, repair or alteration to property).

Civic amenity site — see household waste site.

Co-disposal — the practice of disposing, simultaneously, of household waste and industrial (normally liquid) wastes.

Commercial waste — consists of waste from premises used wholly or mainly for the purposes of a trade or business or the purposes of sport, recreation or entertainment, excluding household and industrial waste, mines and quarry wastes, and agricultural wastes.

Containment site — a landfill site designed to contain the waste and prevent the escape of potential pollutants, for example leachate.

Controlled waste — household, industrial and commercial waste: excludes mine and quarry waste, agricultural waste and (except in certain circumstances) radioactive wastes.

Directive waste — see EC definition in *What is waste?* section pp. 8 – 14

Duty of Care — this is the responsibility placed by Section 34 of the EPA 1990 on all those involved in the production, transportation, storage treatment or disposal of controlled waste. All reasonable measures must be taken to prevent the illegal handling/deposit of waste and ensure that the waste is only transferred to an authorized person (i.e. a registered carrier/the holder of a waste disposal licence) and is accompanied by a written description (transfer note).

Fly-tipping — the illegal deposit of waste, normally on or adjacent to the public highway.

Household waste — consists of waste from a private dwelling. It also includes waste from residential homes, or from premises forming part of a university, school (or other educational establishment), hospital or nursing home.

Household waste site — a place provided by a waste disposal authority, under the Refuse Disposal (Amenity) Act 1978 for the free disposal, by the public, of bulky items of household waste.

Industrial waste — in essence, consists of waste from any factory (within the meaning of the Factories Act 1961) and any premises occupied by a body corporate (excluding waste from any mine or quarry).

Inert waste — material which does not decompose or decomposes only very slowly, and is virtually insoluble in water. It should be differentiated from building/excavation waste in that the latter normally also contains degradable materials such as timber paper and plasterboard.

Landfill — the controlled deposit of waste to land, often in voids left as a result of mineral excavations.

Landfill gas — gas produced as a result of the anaerobic (in the absence of oxygen) decomposition of putrescible material. The gas consists principally of methane and carbon dioxide in the ratio of approximately 60% CH_4, 40% CO_2.

Landraising/landforming — the controlled deposit of waste in naturally occurring voids or on top of virgin or derelict land to produce a new land form.

Leachate — liquid that passes through a landfill and, by doing so, picks up contaminants from the deposited waste.

Liner — a layer of material (either naturally occurring or an artificial membrane) utilized in the base and walls of a landfill site in order to prevent the escape of liquids into the underlying geological strata.

Municipal waste — refuse collected by waste collection authorities (or by contractors working for the authorities); normally composed of household waste from householders and an element of commercial waste from shops and offices for which a charge is made.

Reclamation — the removal of materials/items from the waste stream for the purpose of either reuse or recycling.

Recycling — the act of reprocessing a waste material in order to produce a useable material. A clear distinction should be made between the act of collecting materials and the recycling of a material.

Registered carrier — a person registered with a waste regulation authority (in whose area they have their principal place of business) under the Controlled Waste (Registration of Carriers and Seizure of Vehicles) Regulations for the carriage of controlled waste.

Site licence — see waste management licence.

Special waste — is or may be so difficult, hazardous, toxic or dangerous to dispose of that special provisions may be required for its disposal. Carriage and disposal is governed by the COP (Special Waste Regulations) 1980. Examples include tyres, phenols, solvents, asbestos and any other material contaminated with a special waste.

Transfer note — this must be completed, signed and kept by all parties to the 'transfer' of waste (as required by the Environmental Protection Act 1990

Section 34 and the 1991 (Duty of Care) Regulations and SI 2839). The note must describe the waste, state the quantity, the form of packaging and the origin of the waste, together with details of the time and place of transfer. Transfer notes for the purpose of documenting the movement of special wastes are referred to as consignment notes.

Waste — see EC definition in *What is waste?* section pp. 8 – 14, which supersedes this definition (as defined by the EPA 1990 Section 75(2), this includes

- any substance which constitutes a scrap material or an effluent or other unwanted surplus substance arising from the application of any process

- any substance or article which requires to be disposed of as being broken, worn out, contaminated or otherwise spoiled

but does not include a substance which is an explosive within the meaning of the Explosives Act 1875.

Waste management licence — a document issued by a waste regulation authority and required prior to the commencement of the operation of a waste disposal/treatment/storage facility. The licence contains a number of conditions regulating the operation of the site, including standards which must be met at all times. Certain other parties are exempt, for example charities.

Waste collection authority — the authority responsible for providing a service for the collection of household waste and (if requested) commercial waste, for which a charge may be made. The authority may also collect industrial waste, but must charge for this service. In England, the district/borough councils are the waste collection authorities.

Appendix 2. BS7750, *Specification for Environmental Management Systems*

The British Standards Institution have produced a comprehensive Environmental Management Standard, which is the first of its kind in Europe. This new Standard should provide an excellent basis for both European and international standards and complements BS5750, Q*uality Systems*, and the European Community Eco-Audit Regulation. Companies have long desired a comprehensive environmental management system by which their performance can be assessed. The Standard has been designed to 'enable any organisation to establish an effective management system, as a foundation for both sound environmental performance and participation in environmental auditing schemes'. In addition, the Standard provides guidance on how to manage environmental performance and the establishment of an environmental policy which should enable business to

- comply with environmental legislation

- ensure products/services are produced, packaged, delivered and disposed of in an environmentally acceptable way

- achieve a strategic plan for future investment and growth to reflect market demands on environmental issues.

The Standard specifies the elements of an environmental management system which can be adopted by any type of organization.

The basic implementation stages of the system are as follows.

- **Commitment.** Management must be committed to adopt the Standard.

- **Initial review.** A preparatory review of the present situation is essential to establish a starting point.

- **Establish policy.** This must outline the environmental policy which has to be implemented by the organization. This document forms the basis of the organization's objectives and targets and must be made available to the public.

- **Organization and personnel.** Appropriate management organization and personnel practices to ensure compliance with the Standard and for communications, education and training procedures.

- **Register of regulations and the evaluation and register of effects.** To establish a schedule of regulations to be complied with; procedures for receiving, documenting and responding to communications with interested parties. Also, for establishing procedures for examining and assessing the environmental effects of all activities, products and services of the organization.

- **Objectives and targets.** The setting out of specific objectives and targets at various levels in the organization.

- **Management programme/manual.** The plan and procedures by which the objectives and targets are to be achieved.

- **Operational control and records.** Daily control to ensure compliance.

- **Audits.** Assessment of performance.

- **Reviews.** Regular reviews of system and start cycle again by re-examining policy.

The Standard provides a welcome support to environmental protection legislation and will assist organizations to comply with their responsibilities. It requires that the evaluation of environmental effects should include

- Controlled and uncontrolled emissions to the atmosphere.

- Controlled and uncontrolled discharges to water.

- Solid and other wastes.

- Contamination of land.

- Use of land, water, fuels and energy and other natural resources.

- Noise, odour, dust, vibration and visual impact.

- Effects on specific parts of the environment and eco-systems.

All of which are of particular importance to the waste management industry.
Sector guidance is available for the construction and waste management industries.

Further reading

Legislation and other publications

Environmental Protection Act 1990, HMSO, London, 1990.

Environmental Protection Act 1990, Waste Management, the Duty of Care — a Code of Practice, HMSO, London, 1991.

CSERGE and Warren Spring Laboratory and EFTEL. *Externalities from Landfill and Incineration*, HMSO, London 1993.

Efficient Use of Aggregates and Bulk Construction Material, BRE, Garston.

Department of the Environment. *Making Markets Work for the Environment*, HMSO, London, 1993.

Department of the Environment. *Policy Appraisal in the Environment: a Guide for Government Departments*, HMSO, London, 1991.

Royal Commission on Environmental Pollution, 11th Report. *Managing Waste: the Duty of Care*, HMSO, London, 1985.

Royal Commission on Environmental Pollution, 17th Report. *Incineration of Waste*, HMSO, 1993.

Royal Commission on Environmental Pollution, 19th Report. *Transport and the Environment*, HMSO, London, 1994.

*Environmental Issues in Construction ,*Vols 1 & 2, CIRIA, London, 1993.

Department of the Environment. *Digest of Environmental Protection and Water Statistics*, HMSO, London, published annually.

Department of the Environment and Howard Humphries, *Managing Demolition and Construction Wastes*, HMSO, London, 1994.

Sherwood, P.T. TRL Contractor Report 358. *A Review of the Use of Waste Materials and By-products in Road Construction*, Transport Research Laboratory, Crowthorne, 1994.

CIRIA. *Environmental Handbook for Building and Civil Engineering Projects; Construction Phase*, CIRIA and Thomas Telford, 1994.

Sherwood, P.T. *Alternative Materials in Road Construction: a Guide to the Use of Waste, Recycled Materials and By-products*, Thomas Telford, London, 1995.

Papers

Kermode, B.C.D. *Highway Maintenance — Towards Sustainability,* IHT Conference, 24 September 1993.

Potter, J. and Earland, M. *Recycling Bituminous Road-building Materials,* Transport Reaserch Laboratory, Crowthorne.

Holmes, A.J. Recycling v reconstruction. *Journal of the Institution of Highways and Transportation,* April 1991.

Isles, M.K. Recycling — the 'greening' of blacktop. *Journal of the Institution of Highways and Transportation,* June 1992.

The construction of Drighlington Bypass, through methane generating industrial waste. *Journal of the Institution of Highways and Transportation,* December 1993.

Kershaw, K.R. and McCulloch, A.G. Environment and planning issues. *Proceedings of the Institution of Civil Engineers, Civil Enginering, Channel Tunnel, Part 2: Terinals,* 1993.

Brown, P.T. *History of Remedial Works to Upper Glendevon Dam.* Paper presented to meeting of AME Scottish Divisions, Institution of Civil Engineers, October 1993.

Ferguson, J. Waste from construction and Duty of Care. *Proceedings of the Institution of Civil Engineers, Municipal Engineer,* March 1994.

Schroeder, R.L. The use of recycled materials in highway construction. *Public Roads,* Autumn 1993.

WIDENER UNIVERSITY
WOLFGRAM
LIBRARY
CHESTER, PA